The Unbelievable **TRUE STORY** about Rex and Our Dino Friends!

DINOSAURS CAN'T ROAR

Well, that was **embarrassing!**

WORDS BY
LAYLA BEASON

PICTURES BY
MARIANO EPELBAUM

sourcebooks wonderland

We all love dinosaurs. We love them a lot.

We know what they are. We know what they're not.

But is that true? New science might show

we should probably question all that we know.

We've seen the T. rex colored from green to blue,
or orange and stripy, and some in shirts too!
Some were funny, some scary, some downright cute.
Meet our friend, Rex (please note the round snoot)!

Our doctor of dinos has some thoughts about Rex.

Get ready for change—this is somewhat complex.

Rex should be lower, with his nose to the ground.

Rex won't like this at all, and he makes an odd sound.

We've seen Rex look tall, just straight up and down,

with his head in the sky and his tail on the ground,

but the doctor declares the posture that's best

is his head pointing east and his tail pointing west.

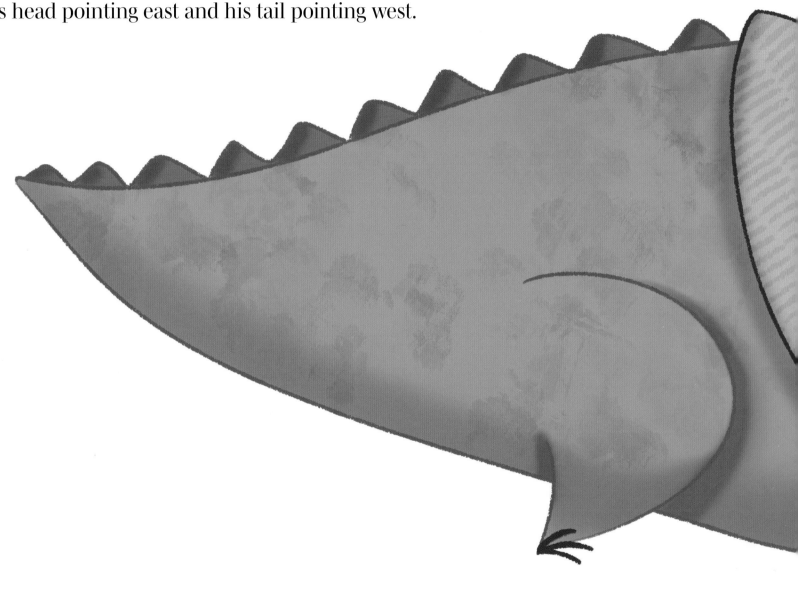

Now let us consider the T. rex's arms—
they're silly and short, not the best of his charms.

Rex was embarrassed his arms were so small.

He couldn't take pictures or catch a ball!

Rex didn't know that his small arms instead

let the muscles grow strong in his neck and his head.

With powerful strength from his back through his jaws,
he didn't need long arms...well, just because.

Yeah, that's right! CHOMP!

Next shall we talk about Rex's real look?

He wasn't the cutie pie seen in this book.

While scientists found Rex indeed had some scales,

he likely had feathers from head down to tail!

Okay, maybe Rex didn't look like a chick.

He had a few feathers, but was still pretty slick.

The feathers were colorful, but not absurd,

as Rex was less reptile, and way more a bird.

DINOS, YOU HAVE TO SEE THIS!

The doc said we're not done. Science found more!
Rex called to his friends to see what was in store.
Triceratops, Stegosaurus, Brontosaurus too!
These dinos are about to learn something new.

Triceratops, famous for horns and a crest,
really wore a bigger and brighter headdress.

But it's not like a feathered peacock, in fact.

To say it's a shield would be much more exact.

A brain in the butt AND a brain in the head?

That theory seems silly, but it's what people said!

That's cool!

$A + B = C$?

$X = Y$?

$E = MC^2$?

Over time, our theories can change. That is true!

Now we know Stegosaurus had one brain, not two!

The Brontosaurus existed, said science in chorus
until they thought, *Nope, that's an Apatosaurus!*

"Bring her on back," the doc started to shout.

"We were right the first time, as it turns out!"

Wherever you just sent me, NEVER do that again!

Though these dinosaurs seem like the perfect friend squad,
if they really hung out, it would've been odd.

Our doc said two friends came millions of years before Rex and Triceratops ever appeared!

There's just one more thing that may awe and amaze
and may change how we think of the dinosaur days.
Get ready for this—surprise is in store!
Three, two, one…yep,

DINOSAURS CAN'T ROAR!

Gasp

NO WAY! Let's show her!

Yes! Dinosaurs can't roar. Current research exists
saying dinosaurs rumbled and dinosaurs hissed,
but the only roars heard anytime, anywhere
came from mammals like lions and tigers and bears.

Roaring or not, let's give dinos three cheers!

They ruled on this planet for millions of years.

We still love them all—Brontosaurus to Rex!

Who knows what we'll learn about dinosaurs next?

Dinosaurs Can't Roar...and More!

Rumble or Roar?

Roaring is a mammal sound made by tigers and wolves, and paleontologists think we assumed dinosaurs could roar because it is such a scary sound—perfect for the great *T. rex*! But new research from 2016 says that dinosaurs—who were ancestors of birds and related to reptiles—probably sounded more like today's alligators and crocodiles. The sound was likely a low, vibrating rumble, something you would feel more than hear.

Is all of this true?

Yes—for now! What we know about dinosaurs (and many other parts of science!) is constantly changing. Scientists always have to keep an open mind, because sometimes they make new discoveries that change what we thought we knew! **Paleontologists**—the doctors who study dinosaurs—learn new things about dinosaurs all the time, so our ideas of what these creatures were like have changed over time too. Who knows what we could learn next!

Tyrannosaurus rex

Posture Problem – *Tyrannosaurus rex* walked leaning forward, with its tail out behind it for balance. But in the early days of dinosaur discovery, *T. rex* was depicted standing upright with its tail on the ground. Movies and cartoons continue to use this incorrect depiction, even though scientists learned about *T. rex*'s correct horizontal posture in the 1960s.

Arm Advantage – *T. rex* has always been teased for its puny arms, but a recent paleontologist says they are the key to *T. rex*'s powerful neck and jaws. Because it had fewer muscles in its arms and more in its head, the dinosaur was able to have a ferocious bite. And compared to humans, *T. rex* would still win at arm wrestling—it could lift around 430 pounds!

Feather Frenzy – Paleontologists in the 1990s discovered dinosaur fossils with feathers, which made them rethink the belief that dinosaurs were entirely scaly like reptiles. Dinosaurs are actually ancestors of modern-day birds—thus, birds ARE dinosaurs. In 2015, the fossils of three dinosaurs related to *T. rex* were found with proof of feathers, so *T. rex* also likely had feathers. Many paleontologists say *T. rex* probably had mostly scales with some feathers down its back.

Triceratops

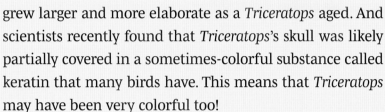

Triceratops is known for its distinct three horns and bone crest. Paleontologists believe that the horns and crest grew larger and more elaborate as a *Triceratops* aged. And scientists recently found that *Triceratops*'s skull was likely partially covered in a sometimes-colorful substance called keratin that many birds have. This means that *Triceratops* may have been very colorful too!

Stegosaurus

In the late 1800s, paleontologists noticed that some dinosaurs had a large cavity in their hips, and someone suggested they may have had a second brain in their backsides. *Stegosaurus* was perfect for this "butt brain" idea because it had a very small brain in its head and could have used the extra brainpower to move its large tail. That turned out to be untrue, though, as we now know. So the butt brain is just a myth!

Brontosaurus

Paleontologists sometimes have a tough job naming different dinosaurs. *Brontosaurus*, the long-necked "thunder lizard," was first named in 1879 by paleontologist Othniel Charles Marsh. Before that, Marsh also named a similar dinosaur, *Apatosaurus*, in 1877. Even though *Brontosaurus* became famous in popular culture, scientists in the early 1900s started to believe the two dinosaurs were related and called both species *Apatosaurus*. But once again, new science showed we should question what we thought we knew! In 2015, paleontologists said a new study shows *Brontosaurus* was different enough from *Apatosaurus* to have its own name. *Brontosaurus* is back!

210 million years ago
First mammals evolve

150 million years ago
First birds appear

65 million years ago
Extinction of most dinosaurs

235 million years ago
First dinosaurs evolve

155–145 million years ago
Stegosaurus and *Brontosaurus* live

130 million years ago
First flowers evolve

68–65 million years ago
Tyrannosaurus rex and *Triceratops* live

TRIASSIC PERIOD
250–200 million years ago

JURASSIC PERIOD
200–145 million years ago

CRETACEOUS PERIOD
145–65 million years ago

Early humans evolve 62 million years later!